THE PRINCESS RANCHER

THE PRINCESS RANCHER

By Kelly Gray Williams

BELLE ISLE BOOKS
www.belleislebooks.com

Copyright 2016 by Kelly Gray Williams. No portion of this book may be reproduced or transmitted in any form whatsoever without prior written permission from the publisher, except in the case of brief quotations published in articles and reviews.

Cover photos by Nathaniel Kidd.

ISBN: 978-1-9399305-6-9
Library of Congress Control Number: 2015951842

Printed in the United States

Published by

www.belleislebooks.com

This book is dedicated to my husband Mike
with love for all of your hard work and determination.
Without you none of this would be possible. To my
Children, Ashley and Hannah, because you are the
brightest Stars in my universe and my greatest joy in Life.

"The two most important days in your life are the day you
are born and the day you find out why."

~ Mark Twain

TABLE OF CONTENTS

Chapter One – My Personal History	1
Chapter Two – The Beginning of a New Life	9
Chapter Three – Becoming a Rancher	18
Chapter Four – The Money Pit	22
Chapter Five – Rebuilding a Farm	27
Chapter Six – The Beginning of our Herd	33
Chapter Seven – Economics	40
Chapter Eight – Farm to Table	44
Chapter Nine – Animal Practices	51
Chapter Ten – The Challenges of Ranch Life	56
Chapter Eleven – Reinvention	65

"Working cowboys are the embodiment of the true American spirit. They live a rugged, clean life: a difficult, yet simple life . . . "
~ Clint Eastwood, foreword, *Gathering Remnants: A Tribute to the Working Cowboy*

THE CODE OF THE WEST

1. Live each day with courage.
2. Take pride in your work.
3. Always finish what you start.
4. Do what has to be done.
5. Be tough, but fair.
6. When you make a promise, keep it.
7. Ride for the brand.
8. Talk less and say more.
9. Remember that some things aren't for sale.
10. Know where to draw the line.

~ From the book *Cowboy Ethics: What Wall Street Can Learn from the Code of the West* by James P. Owen

CHAPTER 1

My Personal History

I wasn't born a rancher's daughter. Cows and horses were not part of my daily routine, but caring for them does rest solidly in my DNA. I come from a family history rich in the traditions and hardships of farming—a history full of lives spent scratching out an existence on a piece of land, and full of people hoping and praying that the land to which they gave their lives could give them some sort of living in return.

The defining moment in my family's history occurred on a windswept dock in Italy. My great-great-grandfather had decided his fortune would be made in America. For a young man living in Florence, Italy, to give up his homeland to travel to an unknown country was a bold move—but a necessary one, in his mind. Opportunities for a young, poor boy living in Italy were scarce, but the options America offered seemed endless. He saw the country as a vast, untamed place perfect for a man such as himself, with wild dreams and a willingness to work hard.

In preparation for his trip to America, he realized he needed a wife. The details are sketchy, but I do know he struck a deal with my great-great-grandmother's father. Arrangements were made, and a few weeks later, before boarding a ship bound for America, two strangers met and married.

As my great-great-grandmother stood on the ship's deck, watching her beloved country slip from sight forever, a burning fury began smoldering in her soul. It was a fury that kept burning until her last breath.

My ancestors arrived safely in America and settled in the Midwest to begin their lives farming. They raised crops and livestock, carving out a meager existence, like all the other farmers during that time in history.

My great-great-grandmother didn't have much that she could control, but she did have one thing: in her utter defiance against a land she loathed, she refused to adopt the language. To the day she died, she spoke not a word of English. For protection, she wore garlic around her neck to ward off the bad spirits that seemed to have taken over her life—a life that had taken her to a strange country with a man she never wanted to marry. It was a life wasted, as far as she was concerned.

The birth of my great-grandmother, Mary, softened the general feeling of gloom that enveloped her mother's existence. Mary was jolly and sweet, and completely responsible for my love of food and cooking. Food was the center of her universe, so whenever we were together, cooking became our way of bonding. Even though she is gone now, I still connect with her every time I cook a meal. Her much-used cookbook sits in a cherished spot on my bookshelf, reminding me of her gentle nature and her infectious laugh. Her spirit is a spirit I share. When I was in my early teens, we drove in her red convertible from Sarasota to Key West to visit her son. She was in her seventies at the time—yet she drove laughing all the way, her white hair flying in the breeze.

From that moment on, I knew I wanted to live my life as she lived hers. Age only defines us in calendar years. A zest for life follows us forever.

My great-grandmother had a lodge in Minnesota with her second husband, a Cherokee Native American whom I called Gramps. One summer when I was ten, she invited me to come and spend a month with her and Gramps and my grandmother at the lodge. It was one of my first vacations without my immediate family, and for a ten-year-old, traveling alone was empowering.

It was a month of pure bliss. Gramps took me hunting and fishing. He taught me how to read signs in nature, how to spot animal trails, and how to understand the different sounds of the woods. Best of all, he told great stories. I would sit quietly at his feet whenever we stopped for a rest, totally immersed in his tales, imagining myself in his world—a world I found exotic and exciting. My grandmother and great-grandmother, for their part, taught me how to cook and sew. On lazy afternoons, we made homemade pasta. Noodles strung on rope all over the kitchen created a colorful maze, like stalagmites in a cave. Messy pots soaked in the sink; savory, bubbling sauce on the stove dared me to steal a taste. My grandmother and great-grandmother told me stories about my ancestors, about the women in our family and their strength and perseverance. I was fascinated and proud.

During one of our trips to town, my grandmother and great-grandmother took me to a fabric store and let me pick out fabric for my new school wardrobe. Once we returned to the lodge, I turned circles on the big dining room table as they measured and prepared. In the evenings, I would curl up with a book as they sewed endlessly into the night. On my last night, as I packed my new fall wardrobe into my suitcase, I marveled at the craftsmanship, the perfect tailoring of my jackets and skirts. It was a month I will never forget.

I am not sure if my grandmother or my great-grandmother ever knew what a huge impact their strong presence made in my life. It was monumental, and I am the person I am today in part because of them. They taught me the art of cooking and what it means to feed the people you love. As I reflect on the past, I realize that my mother, grandmother, and great-grandmother all gave me my passion for entertaining, and the understanding of how that desire is woven into the fabric of our family. Honoring the people we love with a great meal is who we are as women, and that love has shaped my passions to this day.

When I was growing up, my mother often repeated a phrase that irritated me to no end (especially during my adolescence, when most things are irritating anyway). When describing our family and its descendants, she would proudly refer to us as "hearty, peasant stock." The words felt like nails scratching on a blackboard. They made me crazy.

With my limited experience, I failed to understand why I should be proud of this ancestry. My goals in life were infinitely more interesting, loftier than working the land or cleaning up after the animals that walk it. Little did I realize that I could run, but not hide, from my heritage. Like a fish swimming upstream, eventually I would end up back where I started.

I may sound at first like a bit of a snob, but in truth, I was a product of a particular time in history. I entered my formative years during a time of what I like to describe as "importance by excess." I was born in 1960, and by the time I became a snotty teenager, life was all about the bling. The most popular TV shows were *Dallas* and *Dynasty*, both of which featured story lines packed with intrigue, glitz, and characters with bank accounts as big as their shoulder pads. It didn't matter if the people on these shows were good, accomplished people, or if they made their money honestly. Values, integrity, and hard work were for the common folk.

It was a time that my friends and I used to describe as a time of "big cigars and motor cars." It was our moniker for "the rich life." There was a lot of pressure to outdo your neighbors. Many of my colleagues had unlimited corporate spending and expense accounts, and—until the business world became so conservative—they enjoyed indulging in three-martini lunches. Life was for the taking, and the upwardly mobile took it.

I entered adulthood with great expectations and a winner-takes-all mentality. I was going to live big, travel the world, and be important—maybe even break through the glass ceiling. One of my greatest strengths is making things happen. I am extremely good at taking an idea and making it a reality. Sticking with it, however, is a different story. Once I'm done with something, I like to move on to something new. I get bored easily, and as a result I like a constant array of new challenges and changing scenery.

I met my match when I met my first husband. He had plans as big as mine, and wanted it all. He was a visionary. As president of a small, very old Virginia food company, his plan was very clear: he was going to grow the business and take it to the big time. We met during my first year out of college. It was a perfect storm.

I moved to Virginia in 1985. I was twenty-six when we married, and I had my first daughter a year later, and my second daughter two years and some change after that. That was how I did things, a human tornado, raging through life.

God laid out His plan for me, and I lived a good many of my adolescent dreams. With a team of very talented people, company management—a president, vice-president, regional managers and eventually a full corporate board of directors—took our sleepy, hundred-year-old company and turned it into the third-largest food distributor in America.

Performance Food Group went public for the first time in 1993, representing the culmination of a huge dream. The stock would trade on the NASDAQ exchange, and the size of its initial public offering—the first sale of stock by a private company to the public—would be decided in New York, in meetings between company management and the investment bankers hired to take the company public.

It was a heady time, and to say that I was living the American Dream would be a huge understatement. Experiencing firsthand what it is like to take a company public was an amazing journey— something most people never get to do—and it reinforced my belief that if you work hard, anything is possible.

As a national food distributor, we did business with everyone from large, well-known corporations to small startup manufacturers. We dealt with national chain restaurants and chefs of every caliber. Early on, our company bought Fresh Express, an all-inclusive salad producer, from its founder. Bagged salads hit the nation.

Food took me all over the world. I have traveled to so many exotic locales, countries with an array of different cultures and food experiences—some over the top. I have stayed in hotels that were former palaces, ridden the Orient Express, and careened through Switzerland, living like James Bond. My experiences have created the rich tapestry of my life. Every trip is woven into a timeless array of events. Places too numerous to describe are housed in my subconscious, *felt* but imperfectly recalled. I had the privilege of leading a life that most people only dream of.

The greatest part of my adventure by far has been the food. My dining experiences are my most cherished. The creativity it takes to become a great chef is something that endlessly fascinates me. A great meal is an event in itself.

When I was eight months pregnant with my second daughter, I had the chance to visit a culinary school in Ireland. Located in the Irish countryside, the school was housed in an old mansion that was said to be haunted, which in my opinion added another layer of mystery to the ancient estate. A branch of my ancestors came from Ireland, so the chance to see a country that was part of my heritage and experience Irish cooking techniques at the same time was an opportunity I wasn't going to miss. My family thought I was crazy to travel so far from home during the last trimester of my pregnancy, but when I set my mind on something, there is no stopping me.

During my stay, I was able to spend a few days studying with the school's chefs. It was an eye-opening experience. The concept of cooking with what is available and fresh was firmly ingrained in their culinary philosophy, and it made a lasting impression on me.

Since then, I have had many opportunities to experience the cooking of other countries. I marvel at the differences, yet understand that each culture's cuisine speaks to me in their own special ways. Visiting Turkey in the late '90s, I found Istanbul's main spice market to be mind-altering, offering myriad smells and exotic flavors which were so abundant, yet unknown to me and my then-current knowledge of spices. The diverse cuisines of Spain, France, Switzerland, England, Bangkok, and the Caribbean islands are all woven into my thoughts on cooking.

I lived life in a huge way. I had all of the trappings of wealth, yet I didn't feel any more important than I did on the day I'd started out. I experienced a great many triumphs and milestones brought about through the work of others. I lived with the accomplishments, yet wasn't directly responsible for any of them. No busting through the glass ceiling.

I am certainly thankful for the incredible life I had. I lived the big dream, and I will cherish it all until the day I die.

In the end, though, I figured out that the life I was so keen on living was rather hollow.

When I'd stood at the altar as a young woman, saying my vows, the "forever" meant something to me. I didn't take the commitment lightly. When it became obvious that my marriage was melting down, it was devastating. The failure of a relationship is complicated. There is never an easy explanation for it, and it is always filled with pain and regret.

Mid-life hit, and my world came crashing down. I walked away from my life, my marriage, and the easy, self-indulgent cocoon I was living in. I did a complete 180. I was now going to have to do something to take care of myself.

While my children were finishing high school and preparing for college, I went back to school to study herbal medicine and massage therapy. I had always been interested in Chinese medicine and the incredible store of knowledge housed in the teachings of our ancestors, and the study of the integration of mind and body and the ways they interact with natural healing remedies was something I had always wanted to pursue. Now, I took the opportunity.

Since then my life has taken many twists and turns. My beautiful clothes still hang in the closet, alongside my exquisite jewelry and dozens of shoes and handbags which, most days, just collect dust. My new life doesn't require any of these things. I happily traded my designer clothes for a worn pair of blue jeans, battered cowboy boots, and ragged fingernails.

I became the Princess Rancher.

CHAPTER 2

The Beginning of a New Life

I believe that when you are ready to change your life and learn something new, the teacher appears. For me, the teacher became my second husband.

Mike is an avid horseman, a hardworking man who knows a little bit about everything, the type of man that seems to be going the way of the dinosaur. When it comes to making big decisions, he often jokingly says, "Don't ask me, I'm just the handyman." His brother always teases by telling him that when the last fence board is up, he will be kicked to the highway. I respond that I will always make sure there is one board missing.

Destiny is a funny thing. I had been given a brief preview of my future with Mike ten years before I married him, but at the time I'd had no idea what I was seeing.

I'd started riding later than most. Owning a horse had always been on my bucket list, but until I moved to the country, it hadn't been possible. In the summer of 2000, when I was in my early forties, my first husband and I moved to a small town outside Charlottesville in pursuit of a quieter life in the country.

Suddenly, my dream of owning my own horse was within reach.

At the time, I didn't really know a lot about caring for a horse, or how to ride, so I sought the advice of a professional. My daughter's soccer coach happened to be an avid horsewoman, so I asked her for help. She took me to a man she knew whom she felt would be the perfect teacher for a novice like me.

My lessons with this cowboy began in late spring of 2001. He started me out riding bareback on a well-trained horse that he owned. Riding without a saddle helped me develop my sense of balance. It also helped me understand how a horse moves, so that I could learn to ride with the motion of the horse, and not against it. A good rider needs to learn how to be the leader in a horse-rider partnership in order to stay safe and in charge. As herd animals, horses are used to following a leader. As a rider, it is up to you to be that leader—otherwise, your horse will take over, and then you will be at his mercy.

Several months into my lessons, I felt ready to start looking for my own horse. The horse I rode during my lessons was very calm, so I felt my instructor would be able to suggest a horse that would be a good match for my riding ability. I needed a horse that would continue to teach me to be a strong rider as I built my skills and my understanding of horsemanship. He told me he had just the horse: a four-year-old gelding named Cody.

Despite his relatively young age, Cody seemed to be a good match for me: I trusted him, and he seemed to accept my leadership. I also trusted the cowboy's advice, so I purchased the horse from him the following spring.

I'd expected a well-trained animal that would help me in my efforts to further my riding. Instead, I got a wild, headstrong horse that took advantage of my inexperience at every turn. It turns out I paid a lot of money for an accident waiting to happen.

The Beginning of a New Life

A green horse and a green rider is what my husband calls black and blue!

By then, though, it was too late to renege on the deal. The only choice I had was to figure out how to keep the upper hand with my headstrong colt, and learn to ride and train him at the same time. It would not be an easy task.

It would also be several months before the barn I was having built to house my new colt would be completed. While he stayed with the cowboy in the interim, I pushed the man for as much knowledge and help as I could get. My lessons continued, and though I could tell by the look on my instructor's face that he knew what a terrible match Cody and I were, in the end he wanted the money more.

After one of my weekly lessons, the cowboy took me out for a quick sandwich at a little gas station deli near his farm. As we were leaving, he briefly introduced me to a man who was entering the store. It was a two-second introduction, and then it was over. I didn't give it another thought.

Eventually, my barn was finished, and it was time to take Cody home. I am sure the cowboy was relieved—I was a constant reminder of his deception, and even the most unethical don't like to be reminded of their misdeeds. I, in turn, was happy to get away from a man who took advantage of his uninformed clients in the name of money.

Ten years passed. By then, my life was drastically different. I was single, living in a small condo in Charlottesville, with a beautiful view of the Blue Ridge Mountains. A lot of water had passed under the bridge. I still owned my first horse, Cody, and he lived on the farm where I'd lived during my marriage. My ex-husband continued to let me ride and use the facility, but I had a new horse. I had never been able to establish myself as the boss in my relationship with

Cody—I just didn't have the skills to get through his issues and put myself in charge. So he became a pasture ornament, and I purchased the horse I should have gotten right from the start.

One day, out of the blue, a friend of mine called, wanting me to meet someone she knew because she felt that we would be a great match. I agreed to meet everyone for dinner. I figured it would be a fun evening out regardless, and anyway—nothing ventured, nothing gained.

When I met my date, he looked vaguely familiar, but I couldn't place him. As the evening progressed, our friends decided that we were getting along well enough, so they went home to let us continue on alone. As it turned out, he was an avid horseman, so we naturally gravitated to talk of horses. When I started talking about my young colt, he got a funny look on his face and asked if my horse happened to be named Cody. I said yes. He then proceeded to tell me about what happened on the day we'd first met—our brief encounter ten years earlier as I'd left the deli. He remembered everything I had been wearing that day, and I finally understood why he looked so familiar.

The greatest coincidence of all was that Mike and his nephew had been my colt's original owners. They had traded the colt for some round bales of hay—and bad hay, to boot. The painful part was that the cowboy had then turned around and sold the colt to me for $12,000. I now understood just how much of a con man the guy really was!

At this point in the evening I should have been crying, but the entire situation was so absurd I just had to laugh. I'm still laughing, shame on me. I was very naïve about horse dealers. Some lessons are painful and expensive.

We were married one year later. It was almost crazy to think that, ten years earlier, in that brief moment, I had caught

The Beginning of a New Life

a glimpse of my future. If someone had told me then that the stranger I was passing in the doorway would become my second husband, I would have laughed in their face. Life can take some pretty interesting twists and turns. I began my horse life with a con man, but ended up with the real deal.

I believe my husband was virtually born on a horse. Growing up, Mike and his brother rode everything and anything. Their mother always jokes that in order to get through the day, she had to pull the blinds, because she couldn't bear to watch the shenanigans outside. Crazy things happened—accidents of all kinds, judgment gone wrong. One of their more infamous acts was tying a young horse to the porch, only to watch the horse melt down and pull the entire porch off the house. There is never a shortage of horse stories in Mike's family—given the many young horses they've introduced to the saddle, Mike and his brother, Warren, have seen it all.

Mike on his horse Boo

Teaching a young horse how to accept a rider for the first time and begin his training under saddle is called "colt starting." Up until fifteen years ago, starting a young horse didn't involve a lot of preparation. Basically, you hopped on and hoped for the best. Usually the young colt got the best of you, unless you happened to get lucky that day. You learned by getting on and getting through it. Bumps, bruises and broken bones were common. It wasn't the best system, but it was the only system anyone knew—until natural horsemanship came into the mainstream, and a kinder, gentler way emerged for both the horse and the rider.

Riding all those young, crazy horses did teach my husband how to have a solid seat and quick reactions. Over time, his ability to feel when a horse was getting ready to buck, bolt, or get out of control before it happened gave him the upper hand and the ability to diffuse a blow-up before it got the best of him. Learning how to stop a negative behavior before it starts can be a lifesaver.

As a kid, Mike showed horses in just about every discipline, which gave him the vast treasure trove of knowledge and experience that he currently uses in his work. Over time, though, his body took quite a beating, and as a result, he began looking for easier ways to train young horses. Ultimately, his need to protect his health led him to a whole new way of starting horses. Today Mike trains using the methods called "natural horsemanship," a common-sense approach to training based on how a horse thinks, rather than how a human thinks. If you understand how a horse learns and train them based on that premise, you will see quicker, safer results.

In addition to riding horses, Mike had worked as a general contractor for thirty years. He made a decent living, and proved

The Beginning of a New Life

Mike in round pen on his horse leading another horse

that if you work hard, you will prosper and grow. That is, until 2007—when the U.S. housing market imploded with all the subtlety of a bomb dropping out of the sky, and a staggering number of Americans who made their living in the building trades were suddenly unemployed. Work became scarce to nonexistent almost overnight.

In early 2007, we began to see the writing on the wall. The loans were getting crazier, land prices were soaring, and everyone, regardless of their experience, thought they could build a house and sell it. The bankers that Mike dealt with were getting nervous about the housing market in general, and we in turn started getting nervous about the inventory Mike had left on his books. He had seven houses in various stages of completion and was already committed to building two more, and nothing was sold. To protect our future, we decided to embark on an aggressive effort to sell everything as quickly as possible.

By 2008, we were in full-blown crisis. The career Mike had relied on for so many years was drying up before our eyes. Almost overnight, selling houses became near impossible.

The fallout meant that we were going to have to reinvent our lives. Mike's career as a builder was over. We had to ask ourselves how we would continue to make a living in our fifties, in a horrible job market. Both Mike and I had worked for other companies, but our experience was outdated and hard to parlay into new jobs. This was the bad part of our employment history. The good part was that most of our experience came from working for ourselves, so we were very good at taking an idea and building it into a full-blown business. We both felt that we needed to capitalize on our entrepreneurial strengths if we were going to survive.

It was a terrifying time, and very stressful. If you have ever lost a job or worked in an industry that suddenly evaporates, you understand the fear that grips your soul when you lose the security of the job that has always paid your bills, and you suddenly have no clear path forward. The experience shakes you to the core. We no longer live in a world where job security is a given; in many cases, it is a thing of the past. So the larger question becomes: how do you continue on? How do you survive?

In times of uncertainty, you have to be willing to take stock of your strengths and weaknesses, be bold, and forge your own path. This was one time that neither of us could sit back and have a pity party.

After a year of intense soul-searching and just plain stubborn determination, we threw ourselves into the first part of our ranching business, which focused on horse training and colt starting. Since we work all of our cattle on horseback, training horses would be the logical first step in our business plan, with the full cattle operation representing the final phase.

Armed with a lifetime of horse knowledge, we realized from our research that people needed what we had to offer. We realized that due to the growing popularity of the Internet for information-gathering, advertising online would be the most cost-effective way to let people know about our new business. To get the word out, we put together a first-rate website to act as the platform for our horse-training business.

Horse training can be very dangerous. Most people are not equipped to take on such a big challenge and stay safe in the process. My husband does all of the ground training and first rides. He takes the training very seriously, and is constantly learning and perfecting his methods in order to help the horse and himself stay safe. On the other hand, since I started riding later in life, I bring an entirely different skill set to the business. I am a certified massage therapist and have also studied behavior, aromatherapy, and herbal medicine, so I bring a holistic bent to the psychological and physiological parts of the training. I used my core ability to make an idea into a reality, and applied it to our new business venture. That being said, had my husband not had the experience with cattle and horses that he has, this new life would not have been possible.

As with any business, building ours took time. Life became a string of worries and false starts. We bartered, scraped, and worked hard. We eventually built a clientele that got us through the difficult times, but we never stopped looking for land to lease so we could raise cattle. And—finally—our big break came.

CHAPTER 3

Becoming a Rancher

A cattle business does not exist without land, and because of the explosion in prices and development, obtaining land is the most difficult part of starting a ranching operation. It's one thing to say that you want to find land to buy or lease to start a cattle business; it's another story when the land actually becomes available, and you are confronted with the choice of whether to work to make the dream a reality, or to just keep dreaming.

For three long years, our talk was just that—talk. For years, all of the leased land near us had been owned by a family that parceled it into a vast network of farms serving their cattle business. Then one day, everything changed. Cattle prices were running strong, so the family decided to liquidate their entire operation. In 2010, a landowner from out of town contacted us about a beautiful farm a few miles from our main residence. He had gotten our names from the local farmers' co-op's owner, who knew we were looking for land to lease. Almost overnight, the market for farmland had opened up—but we knew it wouldn't last long. Our day of reckoning had arrived. It was now or never—we had to jump in, or watch our dream pass us by.

It is important to understand what we were facing. The market for selling cattle was strong, and that meant it would be costly to buy them—very costly. We didn't have much time to make a decision, and throughout my hand-wringing, my husband was very honest about the risk, the investment, and most importantly, the hard work involved in running a business of this nature and what that would mean for my life. I am a very hard worker, so the amount of hard work involved in the business didn't faze me. Taking risk, on the other hand, was a whole different animal.

To a large degree, I had always been taken care of; security had always been there. I believe that when you grow up this way, it becomes harder to live a financially risky life. I went from my father to my first husband, and except for a short period in my early career, I didn't have to really struggle with money. Resources were there, along with a regular paycheck and a lot of benefits. I had to worry about other things, but money wasn't one of them. As a result, my risk tolerance was about zero. I liked my life, liked the security of knowing that I had a steady flow of money, and didn't want to do anything to jeopardize that security.

When I divorced my first husband, my zero-risk ideology became even stronger. At that point in my life, I had a finite amount of money and no paycheck, so preserving money was paramount. When it came to my nest egg, I was like a vicious guard dog, so the thought of investing a sizable amount of money into a cattle business was as terrifying to me as jumping out of a plane without a parachute.

My husband, however, is a risk-taker. Life hasn't always been easy for him; he has had to work multiple jobs at various times in his life just to get by. When he became a general contractor building spec houses, risk was the name of the game. His tolerance for risk, therefore, is much greater than mine. In many ways, we balance each other out, but when it comes to taking on risk, he has to drag me kicking and screaming the whole way.

It's easy to talk about a business philosophically, especially one that carries a lot of risk; it's quite another to actually write the check. To fully appreciate our unease over starting a cattle business, it is important to understand American history.

Our country began as an agriculture-based society, with roughly ninety percent of its population involved in farming to one degree or another. Around the 1900s, that number had dropped to fifty percent, which is still respectable—but today, less than one percent of the population claims farming as their sole occupation. In 250 years, the United States has changed from a society in which ninety percent of the population farmed or ranched into a society in which one million beef producers have the responsibility of raising food for roughly 314 million people.

This was a seismic shift, with huge implications for the farming and ranching industry. These changes transformed cost containment and productivity into key issues. If a cattle operation is to be profitable, there can be no room for mistakes. Supporting a business model that depends so much on God and nature is never an easy choice. A person must have deep faith and thick skin in order to adapt to quickly changing situations when dealing with live animals.

My favorite little joke to tell is that I now understand why farmers and ranchers are deeply religious: in this line of work, you are constantly praying to God for small favors. *Please, God, make it rain; please, God, make the rain stop; please, God, end this heat wave.* So many variables remain out of a rancher's control that they must take great measures to manage what they can. It is paramount to the survival of their business.

I had done enough research to know that raising cattle would be a difficult career. But at the end of the day, knowing all that might be against us, and the risks involved, my husband and I did it. In 2011, we joined the one percent of the U.S. population that raises beef cattle for a living.

CHAPTER 4

The Money Pit

Virginia winters can be very unpredictable. Rainy, snowy, cold and wet, or dry and warm—it is hard to know what kind of weather you will encounter from one season to the next, and as a rancher, you have to prepare for it all. Weather plays a huge factor in the amount of forage you need to get through the winter season. The colder it is, the more hay you are going to need to keep your animals warm and their bellies full. Good quality forage is crucial to the cows; it keeps their milk flowing.

It is my first winter as a rancher. The sky is heavy. The weather center is predicting rain that will soon turn to ice and snow. I can feel the moisture in the air as the coming storm makes my bones ache. The only thing on our minds is preparing our cows and calves for the difficult days ahead. It is crunch time. Large round bales of hay must be moved into strategic cover areas so our cows have a dry place to ride out the storm. We have newborn calves and older calves all nursing on their mothers, which means our cows need more food than normal. There isn't a second to waste.

The wind picks up as I watch my husband spear two round

bales at a time, one in front of the tractor and one behind. I pull the collar of my jacket up around my face to cut the biting wind at the back of my neck, and open gates as he passes through. We work like this, traveling back and forth transporting hay, for an hour and a half, until Mike feels there will be enough hay available for the cows if we can't travel the three miles from our house to feed our cattle for a day or two. We cover the tractors with heavy blankets to keep the engines from icing over. Finally, we have done all we can to prepare our animals for the harsh weather.

Chilled to the bone, we head home to get our horses ready for the storm. We bring some into the barn, while others will weather the storm in pastures furnished with open, three-sided buildings that they can stand in to get out of the weather. The horses rear and buck as we move them. They feel the dramatic change in the barometric pressure, and that knowledge makes them anxious and unruly. Their behavior makes me uneasy too, because I know they are dangerous at times like these.

After we finish taking care of the horses, I fill their water troughs. If the power goes out during the storm, our water pumps won't operate, but full troughs will give the horses enough water for a few days. While my husband moves our tractor into the barn, I put out extra hay—like cows, horses also eat more to stay warm. After that, we both work to haul a huge stack of firewood into the garage to keep it dry. We finish just as the first snowflakes begin to fall.

As the storm unleashes its angry fury outside, I silently pray that my children and our animals will remain safe and unharmed through the night. It is all I can do. We have done everything possible to prepare. Now it is up to God and nature.

Life as a rancher is not just about animals and land; it is also about how you feed those animals during the winter. My husband

and I have read all the books about rotational grazing, a method of feeding that significantly lessens a rancher's need for rolled hay during the winter by extending the grazing season. We have also read about how to get by without equipment. But when you start a back-breaking business like this in your mid-to-late fifties, you need equipment. The more you can get a machine to do, the longer your body will hold up. *Ka-ching!* Equipment was the beginning of the money pit.

My husband is a real talker, in an endearing way. He would talk to a stump if he thought he could get away with it—but to his credit, this quality has come in handy in our life together. If he didn't talk to everyone, we wouldn't have found our land or half of the equipment and knowledge we currently have, so I don't mean to belittle such an important reason for our success. But at times, I can get very annoyed by it—especially when he's talking about throwing around large sums of money.

By late June, one month after signing the lease on our new property, we were sitting in the John Deere dealership, ready to sign away our lives for hay equipment, and he was chatting up a storm. My head was spinning, and I thought: *Doesn't he realize we're jumping off a cliff for the sake of feeding cows?!* My risk-averse brain couldn't process this madness, and I just wanted him to shut up and let us bury our future in silence.

We were starting our business in late spring, creating an already stressful situation regarding hay for the coming winter. We needed equipment and we needed it yesterday. We should have started making hay a month earlier. If you wait too long, the hay gets overripe standing in the field and doesn't give you the best first cutting, because the weeds start to take over. Time was not on our side. The sense of urgency was suffocating, and on top of that, we were both stressed and grouchy. Add to that the fact that I was still mourning the loss of my easy life, and I was freaking out about money.

In Mike's defense, it was my decision to go forward, not his. In reality, a bit of the "overindulged princess" mentality was still lingering. But during my pity party, the equipment arrived, to the joy of both of us. It was time to go to work.

There's a saying that used equipment is a pain in the ass, and I was about to find out just how true that statement was. It was already early June, so my husband started cutting hay immediately—and almost immediately, the used tractor we had bought stopped working. This major breakdown was only the beginning of the nightmare. The dealer assured us that they had gone over the tractor with a fine-toothed comb, and that they knew the farmer trading it, who said it was a great tractor. Obviously, it *wasn't* a great tractor, so they hauled it back to the dealership and we said, *Good riddance!*

With our tractor out of the picture, we now had hay equipment and nothing to pull it with. We looked everywhere, left no stone unturned—but at that time of year, there wasn't a used tractor for sale on the continent. "Grumpy" didn't begin to sum up our mood.

After a sleepless night, I got up and started crunching numbers. I was not interested in taking on someone else's problems again, but we were still desperate for a tractor—so rather than buying another used one, we decided to explore the idea of buying a *new* tractor. But this time, I told my husband, no small talk. I thought I would hyperventilate when we signed the new set of papers, so I needed his lips available for resuscitation, not blabbing.

Words will never do justice to the problems we had with our equipment that summer. The breakdowns were constant; the repairs and parts cost us a fortune, and in the end, over the course of the next year, we completely rebuilt our hay baler. This was not the sort of outcome I had been banking on. The money was rolling out the door like water during a spring flood. I was almost apoplectic at this point, constantly checking our bank balance and wondering why I had agreed to this madness. My complete inability to handle the stress was making me nuts, and I know in turn it was making my husband even more determined to make it all work. I give him so much credit during this time for dealing with me at all. It was a tough situation and a run of bad luck, but Mike had to go on, and that is exactly what he did, despite my best efforts to make him crazy.

My beloved husband worked day and night, tackling every obstacle put in front of him. At the end of the day, the hay was put up for the winter, and somehow he survived my craziness and the broken equipment.

This is where I talk about God again. I know He was challenging us, each in a different way. We both had growing to do. God was presenting the challenges we needed, and it was painful, but in the end we persevered and made it through.

Our next task was setting up the infrastructure.

CHAPTER 5

Rebuilding a Farm

During my first marriage, I lived down the road from the farm my second husband and I currently lease. On my daily trips into Charlottesville, I would admire the layout of the farm, with its gently rolling hills framed by the Blue Ridge Mountains in the distance. I was curious about the farm's two-hundred-year history.

My house at the time was built on a portion of the original farm. Our property housed the family cemetery, which I restored because it fascinated me. I felt an eerie connection to the land right from the start. Little did I know at the time how much my life was going to change, and how much of that farm I was actually going to see.

By early fall, with the hay debacle behind us, we turned our attention to the actual farm. Back in the day, it had been a thriving dairy farm, but after the patriarch of the family passed away, the farm had sunk slowly into disrepair. We knew it was run-down and that the fences were in terrible shape—in many instances nonexistent—but we didn't realize just how bad things really were.

Our lease agreement took into consideration the cost of fixing the fences, but our labor was free. Five miles of fence had to be fixed, and in order to make it secure, we would have to rebuild almost all of it. To do this, we had to unwind barbwire along long sections of our land, install strong posts—or use trees, if they were available—and then use the claw end of a hammer to pull the wire tight before nailing it to the posts. In order to make a fence strong enough to hold cattle, we had to run six strands of wire around the entire farm, which meant we were actually building about thirty miles of fence.

The only involvement I'd ever had with anything with the word 'barb' in it was Barbie—certainly not barbwire. To work with my husband rebuilding thirty miles of fence over rough terrain defied logic and common sense. I have never met anyone who would want to do this with their partner. I was beginning to understand why I was part of only one percent of the population. It was a long, hot summer and it was hell. What had happened to my summers languishing by the pool, reading a book, and drinking margaritas? This was *The Twilight Zone*. I was sure I was just having a bad dream and would wake up soon.

I might sound like a brat, but I am being honest. It was a huge culture shock for me—not to mention a huge change in my lifestyle. I had some growing up to do, and time wasn't going to wait for me to get it done. Once you get past the fluff, I am a hard worker and a gritty person deep down inside. I knew I had to dispense with the drama; this was the life I'd chosen. We had to get the job finished, so I needed to buck up and get with the program.

In the end, that is what I did. We needed to get the fence work done and the livestock on the property as fast as possible. When you are ranching, each day that goes by without cattle is one more day that you aren't making money. This realization was very motivating.

The first hurdle I had to overcome was learning to work with my husband. It had been so long since my husband had learned to build fences that he couldn't remember what the work might be like for someone who had no clue how to build a barbwire fence in the first place. Working together was ugly in the beginning, and it's amazing that we both survived it. There were times when we weren't kind to each other. My husband has no patience for learning curves. I, in turn, started to tune him out just to make him mad. Eventually, we created a rhythm that worked—most of the time.

The work progress was slow and tedious in the awful heat. I would tie a bandana filled with ice around my forehead to try to stay cool, and it would melt in less than five minutes. The scorching temperatures made the work totally draining. At the end of each day, we were beyond exhausted. I barely had the strength to put dinner on the table before passing out—only to start over again the next day.

In addition to the challenging terrain and the small amount of salvageable fence, there were the cedar trees. These annoying beasts had grown up on both sides of the fence line, and the wire would catch on their branches, making it almost impossible to work. We had no choice: in order to get a clear line to run the wire, we had to cut back the cedars.

Down came the cedar branches, by the thousands—and with them came ticks. Toward the end of the project, I felt like the monkeys you see picking bugs out of each other's fur in the zoo. There are places on your partner where you shouldn't have to look when you are both fully entrenched in middle age. The imagination suits just fine. Unfortunately, in our case, we had to go there. Yuck!

About midway through the fence project, I started getting hostile. I was so sick and tired of the work; it seemed to stretch on endlessly, with no relief in sight. It didn't matter what the problem was; anything seemed to set me off.

My biggest pet peeve, however, was my husband's unwavering attention to detail. Mike's need to string six strands of wire around five miles of fence was pissing me off. We had countless discussions on why we needed six strands. My husband always seemed to build everything to withstand an earthquake or a hurricane. I told him he was acting like we would be fencing in rabbits and giraffes instead of cattle. I didn't understand his need to build a penitentiary for our cattle. From my limited point of view, it was overkill. It was getting on my nerves in a big way, and I made sure he heard about it endlessly.

The most traumatic part of our fence journey occurred on Father's Day. We stopped by the farm that day just long enough for Mike to check on a few things and tweak some fencing. But Mike never knows when enough is enough. Sometimes not knowing when to quit will get you into trouble—and today was one of those days.

The day before, we had cut off one of the corners in the fence line at a diagonal in order to avoid a power line, and Mike was using a piece of rebar to crank the wire tight: just a bit more, and a bit more . . . You get my drift. He didn't require my help, so meanwhile, I was reading a magazine—and anyway, it was Sunday, supposedly a day of rest.

Suddenly, the truck door flew open, and my husband yelled for the first aid kit. The rebar had recoiled, striking him on the tip of the nose. He was lucky that it didn't take out his eye, but nonetheless, it was not a good scene. His nose was gushing blood, and I was frozen—completely ineffective, in shock.

Luckily for him, he reacted quickly, pouring alcohol over the wound and clamping down hard with a cloth. Recovering, I bolted into the driver's seat and took off while Mike sat next to me, trying to staunch the horrendous flow of blood gushing from his nose. As I drove, he tried to tell me he was fine: we could just go home. I gasped at his flippant reaction to such a serious injury. Was he kidding? He needed a doctor, pronto!

I drove fast and furious for thirty miles to the emergency room, where the attending nurse promptly admitted Mike and said they weren't going to touch his nose with a ten-foot pole until they could locate a plastic surgeon. On Father's Day, no less!

During all of the chaos, I realized we'd brought our cattle dog, Wally, with us—which meant that after tucking my husband into the emergency room bed, I had to drive thirty miles back home to drop off our dog, and then thirty back. Ninety miles later, I was back in the emergency room, where they had eventually located a plastic surgeon to do the sutures. My husband has a gigantic threshold for pain, but when the doctor told him point-blank that he was going to have to shove a huge needle up Mike's nose to administer the anesthetic and that the process was going to be terrible, even Mike shook. It's never a good sign when a doctor tells you something is going to be terrible. Their job is to sugarcoat the obvious—so when you get it that straight, you know whatever they do is going to be some kind of painful!

It was a long day, but Mike survived. Not exactly our idea of a perfect Father's Day—but then again, we can't always control the events in our lives.

After eleven months of twelve- to fifteen-hour days, the job we'd set out to do early that summer was complete. Now, when I walk the fence line to check for problems, or ride my horse along the long, winding fencerows, I see my past laid out before me, in sections of fence. I remember this day or that problem, this joke or that tantrum. Each strand bears the imprint of my life, and my husband's life. It is a tangible piece of ourselves that we have left behind, one that will survive beyond us. Despite all of the hard work and the long hours, it has instilled a sense of pride in us to know that we have put a centennial farm back together, and that it has become once again what it was always supposed to be: a thriving, working farm. The fact that I am partly responsible for this is something I never in a million years would have thought I would add to my résumé. I joke all the time that I should get a circular piece of barbwire tattooed around my bicep, in honor of the miles of it I've strung.

    I have traveled light-years since my first marriage and the moments I spent gazing at that farm. I am no longer an innocent bystander. I survived my coming-out party. I paid my dues. After all of our grueling work, watching our first cattle graze instilled a sense of purpose in me that I didn't have before. I now feel like a bona fide rancher.

CHAPTER 6

The Beginning of Our Herd

"The heroism of the working cowboy isn't a joke . . . It isn't something that has been cooked up by an advertising agency, and it isn't something that cheap minds will ever understand. Cowboys are heroic because they exercise human courage on a daily basis. They live with danger. They take chances. They sweat, they bleed, and they burn in the summer and freeze in the winter. They find out how much a mere human can do, and then they do a little more. They reach beyond themselves."

~ John R. Erickson, *Some Babies Grow Up To Be Cowboys*

Our cattle dynasty started with eight purebred Angus cows—each, the good Lord willing, pregnant, and each with a calf at her side—what we in the business call a "three way." We hovered and tended to our small herd like they were our own blood children.

The downside to starting a cattle business during a strong market is the high cost of entry. We paid prices that would make old-timers gasp. Our starting herd was expensive, because we bought cows with quality, proven genetics from a family long attached to the purebred Angus breed. This attention to detail was important to us. We couldn't have asked for a better partner to help us develop our own herd.

Our first roundup was a festive affair. Friends and family turned out with horses in tow to help us celebrate the start of what would become a yearly tradition. A working day is an explosion of the senses. The experience included everything I imagined it would: hot, sweaty horses; barking dogs; grime; mud; skittering calves; noisy cows calling for their young; the smell of a burning fire; and the sound of a sizzling iron as it brands hide. Laughing, yelling, and genial camaraderie added to the chaos and fun.

The first thing I had to learn about working cattle was how they move. It is lesson number one, and until you really understand it,

Mike's brother Warren in round pen with other horses and calves roping

you will do more harm than good, but when you get the hang of moving cattle, it is a beautiful dance between the rider, horse, and the cow. There is nothing more exciting to watch than the slow movement of a herd pushed forward by the subtle momentum created by men and women on horseback.

In my husband's family, everyone helps with the teaching of a "newbie" except the spouse of the one learning. In my case, it was my brother-in-law who advised me, because I knew I would listen to him. My husband, maybe not so much.

In the beginning, the best way to learn how to move cattle is on foot. You will take a beating if you make a mistake—maybe even get charged at—but it is part of the process.

Movement toward the cow's head turns them left or right; movement at the flank drives them forward. If you put too much pressure on a cow by moving too quickly or aggressively, you risk getting run over, and that is how working cattle gets dangerous. Subtlety is the name of the game when you want to influence the direction your cattle will move.

Another important lesson I had to learn was one about "pressure." Cattle, like humans, have a personal space. When you move into that space, you create pressure, and the animal (usually) moves. Too much pressure, though, and the cows take off; too little, and they won't react. For me, learning to read a cow's body language has been the most challenging part of cattle-driving, thanks to my personality. I can be bold and loud sometimes, and move fast, so I had to learn right away how to tone myself down if that was required. As with riding a horse, an understanding of timing and feel are necessary, as the degree to which you should use pressure depends on the situation. I have a long way to go, but every time we work cattle, I get better. My crowning glory, after several years of mistakes, occurred when I moved a small group of calves from a field into a holding pen, and then separated one cow from the group and drove it into another pen. I did it all by myself, on foot—a proud accomplishment.

On horseback, however, rounding up two hundred cattle spread out over one hundred sixty acres requires another set of skills entirely. Things can get unpredictable very quickly. Many times a rider is required to move at high speeds over difficult terrain. This was a completely new concept for me; the only riding I had done up to this point had taken place in a contained arena. To make matters worse, my horse was primarily a show horse, and had spent the majority of his time riding in an arena as well. Riding out in large open spaces, over varied terrain, tasks a horse and rider in ways that are very different from arena work. As a team, we had to learn how to cross obstacles, water, and rocky hills—natural features foreign to my horse and outside my skill set as a rider. It was a real three-ring circus in the beginning.

For me, getting used to riding on all types of terrain was the most unnerving part of cattle driving on horseback, because anything can happen. Our property contains some rough stuff: soupy lowlands, rocky ravines, water, and a lot of groundhog holes. To cross land like this at a clip, you have to trust your horse completely. In the beginning, though, I don't think my horse even trusted himself, let alone my ability to keep us safe. Crossing water was terrifying for him and a real challenge for me, on his back. Our evolution into an experienced team of horse and rider was pretty ugly at times, and not exactly safe for either one of us. During the learning process, my horse had tantrums and bucking meltdowns, and generally acted belligerent whenever he had to hold down an area by himself, without other horses around. During some of his tantrums, he ran me through trees, tearing my clothes; stomped my hat; and annihilated my sunglasses. We suffered cuts, bruises, and scrapes, but I didn't give up, and I wouldn't let him give up. We were going to become a working team if it killed us. Exposure breeds confidence—and we both needed a lot of confidence.

Year by year, the roundups get easier, as over time, I have developed a better understanding of the strategy behind moving an unruly bunch of cattle into a small area. At the same time, my horse is getting better with every outing. It takes time to make a great cow horse; it doesn't happen overnight. Slowly, we have started to work together instead of against each other, and we are learning the language of a true partnership. We aren't finished yet, but we are getting there.

In any roundup, when the herd is contained, the work begins. Usually we start by separating the dry cows—cows without calves—and running them through the "chute," a close-fitting metal cage consisting of a head catch and a body squeeze that restricts a cow's movement, holding it in place while handlers administer vaccinations and worming. Once the dry cows have been doctored and turned back out, the fun begins.

Try to imagine how you would feel if you suddenly became separated from your small child in a grocery store, airport, or shopping mall. Pure, unadulterated panic and terror set in, and as a parent, all that matters is recovering your child. Now multiply that panic by 1400 pounds, as cows try to stay with their calves during the separation process.

A lot of mayhem ensues: cows charging; calves skittering in all directions, horses snorting, running, and jumping as everyone quickly tries to move the cows into one pen and the calves into another. We try to run the pen quietly, keeping it as stress-free as possible, but despite our best intentions, things can get pretty wild. The presence of horses, riders, our dog, and a bunch of people on foot really ramps up the energy. I live for the excitement of it all. I have a lot of wild stories related to working cattle—perfect campfire stories after a long working day, stories that ultimately grow in size and stature over time, creating a great cowboy tall tale.

Once all the calves are separated from the cows, the bull calves are roped, castrated, vaccinated, and wormed. The calves that are too small for weaning go back to their mothers after they are run through the chute for their shots and worming; then, when we are finished with the cows, they all pair up again and are turned out with the dry cows. The calves left are the ones that will be weaned; they are doctored, loaded into trucks, and then transported back to our house three miles away for a process that we call "backgrounding."

Our home has one designated pasture where the calves are kept. There, they are fed daily, which helps to relieve the stress of weaning. They get all the grass and fresh water they can consume, and as an added bonus, they get used to people, equipment, dogs, horses—all of the things that happen during a normal workday. This desensitization helps them to be less skittish, which makes them easier to handle once they are turned out again or sold. My husband and I firmly believe in this process. In our opinion, it creates healthy, happy calves that are ready to thrive. In any successful operation, if the animals you are turning back out on pasture are not healthy and stress-free, they won't gain weight like a prepared calf, which means lost revenue.

When I look back at my first days working the chute, I laugh at how wet behind the ears I really was. My first attempt at giving shots was a joke. You have to be quick and know where to inject the animal, and I didn't have a clue. Worming was another disaster. Half the time I would miss the animal and hit my husband instead—boy, did that make him go ballistic, and rightfully so: cattle wormer is a chemical that you don't want on your skin, due to its human toxicity. My oldest daughter, a veterinarian, chastised me for not wearing gloves or taking the procedure seriously.

Today the area around the working chute looks completely different. I have a long table set up with medical supplies packed in containers. Needles are labeled by size and placed in containers. Our medicine guns are loaded and ready. Wipes, towels, surgical gloves, and wormer are close at hand for ease of use. Animal tags, buttons, and a logbook finish the ensemble—a far cry from my first working day. I'm proud to say I learned my lessons. This is what the learning curve is all about. Experience is making me a better ranch hand.

I have a sense of pride in my work now. It is hard work, filled with danger and its own set of challenges, but I believe that what we do is vital to the survival of our nation. To me, feeding people is a noble pursuit. The day-to-day work has its hurdles, but we persevere, because we are passionate about our business.

When I was a teen, I wanted a big life. I wanted to be wealthy, and I was—but in a shallow, unfulfilling sort of way. Now I *really* have a big life, a very wealthy life—in an important, spiritual way. Learning to let go of the need for material things was my life's lesson, and I am certainly relieved that I learned it before my time on Earth is over, because now I can experience each day in a different, more meaningful way. I have learned that happiness is an ongoing journey, and that many of the things that I might worry about today won't bother me tomorrow. Living in the moment puts the struggles of each day into a more positive perspective. Peace has settled into my psyche, making life richer because of it. My past life paved the way for this life, and for that I am eternally grateful.

CHAPTER 7

Economics

There is a reason why only one percent of the population—or less—does what my husband and I do, and it is the hardest part to talk about: money. It keeps me awake at night, and has added a furrow to my brow that even Botox couldn't fix. It makes me feel like a nag, and it is also the reason why so many people can't survive in this occupation. I was completely unprepared for the daunting amount of capital it takes to start this type of business. The ranching life is a numbers game that is not for those who are faint-hearted when it comes to money.

According to the 2012 Census of Agriculture, Virginia farmers, on average, are 59.5 years old—one year older than the nation's average. Many have either inherited their land and their operation, or have had a lucrative career in another field and are now pursuing a second act. Many can only do this because they have the money and the resources to be able to afford the land prices and build the infrastructure necessary to start a business of this magnitude. Resources aside, starting a ranching business is still like entering a hamster wheel, running constantly in a never-ending quest to stay one step ahead of your bills. A friend of mine is an accountant for a large farming operation, and whenever we get together we always seem to gravitate to talk of money and what we are up against. He usually just shakes his head and talks about the amount of debt that large farming operations have to carry, and admits that he seriously doesn't know how any of them sleep at night. The truth is, in our business, most ranchers don't sleep if they start thinking about money.

Today, the typical American cattleman has around twenty head or fewer in his operation. What does that number mean? It means that to make a living, most of these people are doing something in addition to ranching—or they are retired, content just to supplement their income.

Currently, our operation sustains about ninety producing cows. The rest of our animals are calves, which puts our total number around one hundred and eighty head. The National Cattleman's Association classifies us as a large private operation—compared to the national average—but from a numbers and profit point of view, we aren't nearly large enough.

To be able to put some money into our own pockets, my husband and I need more producing cows. Our goal is to reach one hundred producing cows over the next few years. The rub is that while we need large numbers in order to make a profit beyond just paying our bills, this in turn means more investment. This is precisely why most ranching is done on a part-time basis: pretty soon, you are up to your eyeballs in capital investment, and all of your inventory has four legs. They can die on you at any time, and then you're out of luck—and business! This is the most terrifying part of all, so you have to be mentally able to handle the risk.

One of the reasons a livestock operation like ours is so costly is because the cows never go away. They need to be taken care of three hundred and sixty-five days a year. On the other hand, once the calves are large enough to be weaned, they can be sold. Therefore, the main bulk of our expenses are fixed.

I have run every number a million times, and I know that the right number of cattle will create the necessary profit for us without adding exponentially to the overall cost of the operation. Many of our costs are fixed costs that will actually decrease when distributed over more cattle. The problem becomes *labor*.

My husband and I run our entire operation without pay. In order to make our business profitable, we need to be able to continue to do the work ourselves. Many old, established ranches out west have added other types of businesses to their operations in order to survive, because even though cattle prices are at an all-time high, so are most ranchers' costs. We run our business as lean and mean as possible, but for someone as risk-averse as I am, the numbers are still the hardest part for me to wrap my brain around. We barter, trade, and sell equipment we don't need in order to get the tools we need to survive.

However, there are elements paramount to a strong cattle operation and breeding program that you can't obtain through bartering. One of most important is a comprehensive animal healthcare program. When you raise livestock for a living, you can't cut corners when it comes to illness prevention. An effective healthcare program consists of immunizations, fly and lice control, quality forage, clean water, and a top-notch mineral program. The last of these provides our cattle with all the essential vitamins and minerals necessary to keep them healthy—much like a human multivitamin. Minerals can come in solid fifty-pound tubs, or as a liquid, in a tank called a lick tank. In our operation, we use mineral tubs instead of a lick tank, because we believe they are kinder from a land management point of view. Instead of a large muddy mess around a lick tank, we opt for strategically placed tubs scattered all over the farm to encourage herd movement. This practice is our unique way of achieving rotational grazing without the infrastructure costs associated with interior fencing—and on top of that, the tubs have a lot of positive benefits relating to animal health. The negative aspect is cost. Each tub costs roughly forty-five to fifty dollars, with a monthly cost that runs around one thousand dollars or more, depending on the season. The price is hard to swallow, but the benefits far outweigh the costs—having a strong health maintenance program is like having animal health insurance, and it is one area where we won't skimp.

Our next-biggest expense is the cost of gasoline. We can't run our operation without it, so we feel the pain at the pump more than the average consumer. Land maintenance costs, land leases, and equipment maintenance add to the rising figures. Once we figure in medicine, feed, wormer, and lice control, the monthly cost is nothing to sneeze at. But these are all items ranchers can't ignore, so we have to find a way to pay for them.

The list expands from there. I've just touched on some of the main expenses involved, but they should illustrate why raising cattle is not an easy way to make money. There has been more than one sleepless night in my household, but somehow we survive, and grow. I'm proud of how far we've come in just a few short years, even if it has been economically challenging.

CHAPTER 8

Farm to Table

Traveling years ago in Vienna, I marveled at the vast array of fresh foods available in the open-air markets. The colors were vibrant, the atmosphere enticing. Live musicians serenaded customers; champagne flowed around crowded, noisy tables as local city-dwellers sampled handcrafted cheeses, fresh fruits, vegetables, and meats. The offerings were a glorious bounty that tantalized the senses. This type of market has been a way of life in European countries for centuries.

Finally, we are seeing the same trend here in the United States.

Americans are an aging population—and thanks to our poor eating habits, we are fast becoming a very unhealthy, overweight country. Our bad habits have caught up with us, whether we like it or not. The country is currently in the throes of this realization—and when that happens across a wide spectrum of society, self-preservation kicks in. When you have stared your longevity—or lack thereof—in the face, you begin to search for ways to change the game.

The farmers' markets change that game, offering customers options for creating healthier menus. The movement started gaining momentum in the late nineties, when the grocery store chains began to advertise: "Buy Organic; Buy Local." Restaurant chefs jumped on the bandwagon, and suddenly, farm-to-table dining was in full swing.

The media has certainly fueled the farm-to-table fire, with all of the horror stories relating to food that have been published over the past few years. With no idea what really happens on the large cooperative farms and feedlots, people no longer trust large producers. What is unknown is fearsome, and when there is no connection between consumers and food producers, people will be skeptical, regardless of the truth. Unfortunately, contamination does occur in our food supply despite all of the safeguards that are put into place, and as a result, people can get sick or even die—so everyone involved in the industry must continue to make food safety a priority.

My experience in the food business over the past thirty years has given me a unique understanding of the reality of large-scale food production, and for the most part, I believe that the industry has been given a bad rap. Still, it is impossible to deny the opportunity that this "perfect storm" of public opinion has created for ranchers like me. With the rise of local farmers' markets, this country has experienced a huge wave of consumer demand for fresh, locally grown foods free of hormones and pesticides—and in the wake of this movement, an entirely new sales avenue has emerged for many ranchers.

The greatest benefit of the public movement toward fresh, locally grown food is that it has single-handedly reconnected farmers and ranchers with the consumers in their own communities. This reconnection, and the endless opportunities it creates for farmers and ranchers, fuels my creative spirit in a way I haven't experienced in years. My own time spent in our town's market this past summer lit me on fire. One of the most interesting aspects of working the market was being able to talk to real people about what my husband and I do and how we raise our animals. Generally, I have found people to be very curious about our process, as they seek information that will allow them to trust our brand.

Once you have your customer's trust, you have their loyalty—but you can destroy it with deception. People buy in the farmers' market because they want fresh, local food that hasn't been treated with pesticides, hormones, or antibiotics, and that is what they should get. Unfortunately, where there is opportunity, there is corruption, and for some people, the desire to make money can override the commitment to ethical behavior.

One example of this is the reseller, who passes off produce bought from an auction as locally grown. Often, the produce is not even purchased in the same state where it is sold. Resellers' tactics mislead consumers into thinking they are buying fresh, local food when they are not, cheapening the whole concept of the farmers' market. In order to guard against theses resellers, markets must insist on screening their farmers to be sure that they are really the ones growing, raising, or making the food they are selling.

Most people don't think of the East Coast as having cowboys and cowgirls—those seem to be synonymous with the West—so people love to see our working pictures. I like to emphasize to our customers that working animals on horseback creates a very low-stress environment for the cattle, in comparison to working cattle from ATVs or pickup trucks. This type of environment a rancher maintains for their cattle is important, because animals that are not stressed are generally happier and release fewer hormones—which creates a much better flavor profile for their beef.

What this means for the consumer is exceptional taste. Raised in the Midwest, I grew up understanding what great-tasting beef is—and in my opinion, there is nothing more disappointing on this Earth than a tough, bland steak. Taste is everything, and putting out a superior-tasting product means customer satisfaction, which translates into customer loyalty. Producing a healthy, high-quality, great-tasting cut of beef is the number one goal of our operation. I am still learning the fine balance between giving your customers the healthiest piece of meat you can without sacrificing great taste—because, although my husband and I focus on producing healthy food, we will not sacrifice flavor in any way.

To achieve superior flavor, you have to experiment and find out what works for you. Every rancher has a different set of goals, and therefore different requirements for caring for their cattle. There are many different methods of "finishing" cattle—that is, getting an animal to slaughter weight, which can vary from 900 to 1400 pounds depending on what it is fed in the final stage. Mike and I constantly experiment with feed combinations as we strive to produce beef that is healthy and flavorful. To offer meat that is of a high quality in both aspects, we need to raise lean animals—but at the same time, the animals must have enough fat to allow for marbling, which gives their meat its intense flavor and tenderness.

Eventually, we came up with a system that works for us. We raise our cattle mostly on grass, free-range. When they reach about 800 pounds, we move the cattle into a smaller paddock and feed them a combination of grass and grain until they reach 1000 to 1100 pounds. We use a high-quality grain without fillers or additives, and limit the cows' grazing area—too much movement creates leaner muscle tissue, which makes the beef less tender. By using a smaller grazing area, we create the ultimate in marbling and flavor.

The time of year in which an animal is slaughtered will also affect its flavor profile. During the spring, wild onions proliferate in the cattle's forage, which changes the taste of the meat, making it gamier. This problem doesn't occur in the fall, so we try to get our animals to the right weight for finishing in late summer, and slaughter them in late winter.

Ranching is hard work, but it has one exceptional perk: a freezer full of great-tasting beef. The reward for all of our time and effort comes when our beef hits our forks.

My years in the food industry have given me a passionate appreciation for a great meal. I have sat in some of the finest restaurants in the world, and eaten some of the best beef this country has produced—but I never understood as intimately as I do now what it takes to put meat on a person's plate. Raising beef cattle has been the most eye-opening experience of my life. The work can be dangerous at times, tiring, and downright frustrating, but my approach to food has changed because of it. I now focus on flavor. When you cook with fresh ingredients, you don't need a lot of preparation or sauces to make a meal tasty. High-quality spices, oils, great cuts of beef, fresh vegetables, and herbs all add an intense flavor profile to any meal, creating big flavor without the fuss.

I cook great meals to honor the people I love. When we started having our roundups, I realized how important help is to the success of our operation. My husband and I would never be able to bring in all of our cows and calves, doctor them, and load them into our cattle trailer without help. Our business counts on the generosity of our family and friends. Our working days are long, exhausting, and full of aches and pains, so I say "thank you" to everyone for a job well done by providing them with a roaring fire, drinks, and a full belly at the end of the day.

In many large ranching operations, a chef at a chuck wagon takes care of the meals. I don't have that luxury. I have to plan meals well in advance, prepare portable food for lunch, and finish with a slow-cooking crock-pot meal or grilling for dinner, because I am working all day like everyone else. Meals made in advance make it easier to get food on the table quickly once I get home, when everyone is tired and hungry.

When cooking for a group of ranchers, I take to heart the words of the bumper sticker on my husband's truck: "Eat Beef— the West wasn't won on salad." I shoot for healthy, hardy, real food that isn't fussy—satisfying, stick-to-your-ribs meals that warm the belly and soothe the spirit after a long, hard day in the saddle. The night before a roundup, I typically make all of the side dishes for dinner and lunch. Then I make a list of everything that I will need the following day, so that I don't forget any necessary items: cattle medicine, wormer, cattle tags, paper, pens, food, drinks, utensils, saddles, and horse tack.

The morning of the roundup starts around six a.m. I get the coffee started; then, the final meal assembly begins. I pack lunch in a cooler, load drinks in another, and start packing our trailer with all the essentials for the day: medicine, wormer, cattle sticks, portable folding tables, and tack. Once I know we are set for the day, I put the final touches on dinner.

My claim to fame is my Cowboy Chili, a dish that presents the perfect blend of spicy and sweet flavors. The use of cinnamon adds an exotic twist to the dish, and multiple meats create a filling meal.

Cowboy Chili

Ingredients

- 1 medium onion, chopped and sautéed in olive oil until translucent
- 1-2 lbs. all-natural ground beef, browned and drained
- 2-3 spicy Italian sausages, grilled and cut into 1-inch rounds
- 2 14.5-oz. cans Hunt's diced spicy red pepper tomatoes
- 1 28-oz. can maple bacon-cured baked beans
- 2 Tbsp. brown sugar
- 2 Tbsp. cinnamon
- 1 Tbsp. chili powder
- 1 Tbsp. garlic powder
- 1 Tbsp. onion powder
- 1 Tbsp. salt
- 1 tsp. dry, ground ginger
- 1 tsp. Hungarian paprika
- 1 tsp. Sriracha sauce

Combine all ingredients in a slow cooker and cook on low for a minimum of two hours. Garnish with cheese, sour cream, and green onions. Serves 8-10.

CHAPTER 9

Animal Practices

I cannot write a book about being a rancher without talking about animal rights. It is a passionate subject for many, so I feel it is important to talk about it from a rancher's point of view.

I will preface my comments with the acknowledgment that there are always exceptions to any rule, but from my personal experiences with fellow cattlemen, I don't know anyone in this business who would intentionally mistreat his livestock. If an operation abused its cattle, it would inevitably lose animals—and in this industry, your animals are your capital, so ranchers who invested a large amount of money in their cattle have strong incentive to treat those cattle well. Indeed, many operations have their entire family fortune wrapped up in their business, and in many of the larger operations, this represents millions of dollars at stake, with several generations of family members depending on the success of the business. The high stakes are nothing to take lightly. It's stressful enough to have your money tied up in inventory that is alive, much less intentionally abusing it.

But even though a rancher would never mistreat their animals, working with the cattle can still be a difficult task. It doesn't matter what sector of the cattle business you are in—at some point you are going to have to handle your animals, and if you don't have a proper way of doing that, someone is going to get hurt. Ranching is a tough, dangerous business, but most operations strive to streamline their process as much as possible in order to eliminate potential injury to people or animals.

In the world of livestock, Temple Grandin, professor at Colorado State University and famed doctor of animal science, has changed how the industry works with cattle. Grandin attributes her success as a livestock facility designer to her personal struggle with autism, a neurological disorder that creates a hypersensitivity to noise and other sensory stimuli. Like many other animals, cattle share this sensitivity, and can react strongly to even small changes. By taking the animals' sensitivities into consideration, designers of cattle flow systems like Dr. Grandin can create programs that minimize animal stress.

Over the last few decades, the professor's heightened visualization skills have helped the cattle industry design more humane animal-handling equipment. Her work has had a tremendous effect on how large numbers of cattle are handled in feedlot operations, and her research has helped pen manufacturers design better layouts for pen and chute systems. For smaller, family ranching operations, Grandin's studies have helped solve the problem of how to set up a plan for doctoring and separating cattle, by providing ranchers with a system that relies on cattle's natural flow patterns. Capitalizing on this natural movement when working with cattle helps create the least amount of stress and injury for both animal and handler alike.

In our operation, my husband constantly tweaks our system to filter cattle into our head chute in a manner that helps us avoid injury. It can be an imperfect process, and it improves with time as we make adjustments, but at least for now, we have created a system that works for each set of circumstances we encounter. The bottom line is: we care about the well-being of our family, our friends, and our livestock.

A smaller operation typically has limited resources, which means that paying extra attention to how it handles its stock will directly affect the operation's bottom line. Family ranches know each animal personally from birth to sale. This intimate knowledge goes far in relieving the anxiety some customers have about animal welfare.

However, even for the best ranchers, dangers exist that are totally outside of our control. Freak weather and unpredictable acts of nature create tough situations for ranchers, who endure all sorts of natural disasters, from floods and fires to raging snowstorms and disease. And unfortunately, on occasion, death is also a part of this natural process.

My first experience with death occurred in March of 2012, during our second year of business. After a long winter, spring had finally begun to creep into our part of Virginia. A light breeze tickled my cheeks as I conducted my daily check on our pregnant cows. I could barely contain my excitement as I waited for the day when I would see a few tiny calves dotting the landscape like oversized cow pies.

In mid-March, that day came. Calves the size of our dog Wally appeared, bringing sheer joy in their wake. It was an idyllic scene—until one day, the dark side of nature reared its ugly head.

The hunting season that year was not a good one. The locals were particularly messy, not bothering to properly dispose of their deer carcasses. Many were strewn carelessly about. This laziness drew in buzzards. These gigantic birds were everywhere, perched in the trees, salivating, ominously waiting for a fresh meal.

As soon as the smell of blood from the birth of a newborn filled their nostrils, the buzzards' hunt began. Down they swooped. Their goal was to catch a calf at its most vulnerable: directly after its birth, before it is up and moving. Once a calf starts moving and nursing, its odds for survival are much better.

Not all of our calves were lucky. Not every cow could protect her young. When an opening presented itself, the buzzards torpedoed in and, with fierce determination, pecked out the eyes and anus of the newborn calf, ending its life immediately.

Finding an animal dead is one of the worst feelings I have ever experienced. It is heartbreaking and sad when all you can do is look on in a helpless state of pain.

On this particular March morning, Mike and I found two pitiful lumps in the field. The mothers of the dead calves stood watch over the bodies.

For the safety of the other calves yet to be born, it was imperative that I remove the dead calves as quickly as possible, in order to get the hungry birds moving away from other potential targets. One at a time, I carried the forty-five-pound calves. Trudging through the remnants of that year's final snow, I trekked deep into the woods, as far away from our other pregnant cows and newborn calves as I could physically get. I was winded and sad, and tears stung my eyes as, with a heavy heart, I watched blood drip down onto my boots. Staring into the dead calves' vacuous eyes, I struggled down a steep hill to find a peaceful resting place for them. I couldn't help it—I was angry. The rational side of my brain told me that this was part of the cycle of nature, but that left little comfort. I prayed I wouldn't have to witness this horror again.

Death, it turned out, was the hardest part of ranching for me so far. My personal experience with it showed me how cruel nature can be at times, and made me feel helpless.

My first brush with death has long since passed, but I will never forget what happens when nature tries to set a balance. Nature has a bright side and a dark side. The dark side lurks deep in the shadows, hidden, waiting for the helpless to appear, a predator's meal, reminding me that the beauty in nature can be fleeting.

Death is part of the ranching life, but my husband and I try to minimize it whenever possible. We can't completely eradicate it, but we can work hard to avoid it. I wouldn't say that death hardens you, or makes you insensitive; I just think you get better at enduring it. Over time I have learned to soldier on, sad but no longer angry. Compassion and caring make us human. The ranchers I know are loving, hardworking people who strive to use ethical practices as part of their business model. Treating animals with respect, and giving them the best life possible, is our personal creed.

CHAPTER 10

The Challenges of Ranch Life

Life in America is unlike life anywhere else in the world. On the positive side, we as a nation cherish the idea that anyone, regardless of his or her circumstances, can realize their dreams and rise to incredible heights. On the negative side, our society may be accused of being too attached to "things."

My own personal journey has taught me how unimportant material things are—but if I said that I didn't love beautiful clothes, a nice house, or a meal out, I would be lying. So how does a material girl find a balance between what she is used to and what she can happily live without? The process for me has not been an easy one, but nature seems to help a lot. It calms the mind and settles the spirit, and best of all, it's free.

A day spent riding with my favorite horse really puts things into perspective. In my heart, there lies the real passion, nature, my horse, and the land I have come to love. There isn't a vacation, restaurant meal, or shopping trip in the world that could compare to this. Working with animals seems to make up for all that is negative in this world—but sadly, as more and more people

The Challenges of Ranch Life

Kelly and Shorty in a large field

move into crowded, urban areas, fewer and fewer people get to experience this peace. The ranching life weeds out the weak and satisfies my mothering instinct. The hard work feels good at the end of the day, and the solitude is great for my soul. I consider myself truly blessed to live this life.

Out West, cowboys and cowgirls live lives that are much different from mine. They often work in such remote areas that they don't even see another person for months at a time. They work in harsh conditions, toil in the cold, brave many predators, and give up most of the creature comforts in life, all in the name of their passion for the ranching life. That's the life of a working cowboy or cowgirl. The pay is minimal, the work is hard, and the isolation can be suffocating at times, but it is what they do, and without them, you wouldn't enjoy that juicy steak in your favorite restaurant or see a grocery store case filled with beef.

My life isn't anything like the life I have described above, but as a rancher, I still have to work hard, and in conditions that aren't always fun. My family's livelihood is based on our animals, so we must take care of them regardless of how we feel that day or what the weather is like. Daily care is all part of the job.

Owning a large herd of cattle and five horses dictates that we "divide and conquer" on a regular basis, in order to get everything done and everyone fed. In addition to our daily chores, strategic planning plays an integral part in our business: we have a short-range plan based on our current feed requirements, necessary farm maintenance, and required improvements to the quality of our pastures, and a long-range plan that details our goals for future growth and the steps we intend to take to meet those goals. Because our operating capital is limited, it is especially important for a family business like ours to plan, and then evaluate that plan on a regular basis—more important than it would be for a large commercial business. Planning requires a complete understanding of our expenses, a record of the number of cattle we are feeding, and an allowance for the unexpected expenses that crop up during the course of a year. We also have to have an emergency plan ready, in case we experience drought or harsh weather that requires more hay than we currently have available. Consistently evaluating outflow helps to determine where changes can be made. Technology and innovation also help us tweak our operation to fully optimize our resources.

When we started our ranching business in 2010, my husband and I did so by taking advantage of the only opportunity we had at the time: cobbling together leased properties and rebuilding farms we would never own. It certainly wasn't the ideal way to begin, but it was the only option open to us. In the cattle business, land is king, and it is hard to come by. In many instances, it is also severely overpriced—so if you want to be a rancher and you weren't born into it, high start-up costs are the price you pay for entry.

The Challenges of Ranch Life

At the end of our third full year in business, I can say honestly that I am extremely proud of what we have accomplished in just a few short years. If there is any area that could be improved, it is on the expense side.

At the end of our fourth year in business, my husband and I began putting together a long-term plan for the future. We decided that in order to continue to expand, and maximize our economies of scale—the savings in our average operating costs that we gain from increasing the number of cattle we own—we needed to live and work in one place, instead of three. Owning our own working ranch would allow us to be more time- and cost-efficient. In addition, we would be able to execute a part of our business plan that required closer monitoring: heifer retention, the practice of retaining quality young cows to put them into breeding service. Retaining young heifers rather than selling them is one way to build a herd, but breeding young cows requires a good deal of hands-on attention. You must make sure that the cow is given the right nutrition for a successful first pregnancy, and that you have selected the right breeding bull, with genetics that tend to lower birth weight so that the first calving is successful.

With these dreams in mind, we decided it was time to sell our house and look for a ranch that would have enough land for expansion. For our personal operation, this was the most sensible plan going forward, because the cost of the leases, upkeep, and taxes on land that belongs to someone else just doesn't make sense anymore with regard to our bottom line. It would be the smartest move we could make for the long-term.

But in order to execute this plan and move on to the next step in our ranching business, we had to seek a loan from a bank. In the early days, farmers and ranchers had a personal relationship, even friendship, with their bankers. Most bankers back then understood what an integral part they played in the lifelong survival of their clients' businesses. Deals were sealed with a handshake. Today, it is just as important to have a solid, trusting relationship with a local banker, because the economic climate has changed dramatically. The banking industry is well aware of the risks involved in farming and ranching. It is also regrouping after the financial meltdown in 2008, so banks are very conservative now when it comes to lending money. The passage of the Dodd–Frank Wall Street Reform and Consumer Protection Act has also tied their hands to a certain degree, unless they can make a strong case for lending to a particular person or operation.

Unfortunately, I am not so certain that today's banking environment is all that friendly toward or understanding of the farmer or rancher. It is a challenge to borrow money when your business doesn't look very good on paper. Gaining the trust of a banker who understands your particular business is virtually the only way to expand when you run a family operation, because they are the only ones who will lend you the money you need, aside from the government, which requires that you first qualify for the loan. Even then, government loans are not a given.

There is nothing wrong with large corporate cattle operations fueled by big bank accounts. Ultimately, they have their place, providing large-scale commodity beef at the lowest prices possible for the average consumer. In contrast, a family-run business such as ours is more cash-strapped and vulnerable. However, on the positive side, our business style is very hands-on, and as a result, we can switch gears quickly to take advantage of opportunities when they present themselves, something a large operation might not be able to do.

A small rancher can also find a niche, and be successful, by focusing on flavor. It is true that when the consumer buys from the smaller producer, they will pay more, but when the quality matches the higher price, it is easily justified. This gives us a way of increasing our bottom line—something commercial operations can't do. Having a specialty market niche may be all a small operation needs to compete within the larger industry, and create a profit base that will help them to achieve financial security in a tough banking environment.

Large or small, all cattle producers face the same challenges, whether on a greater or lesser scale. Ranching is a difficult industry, and it has always taken strength and perseverance for ranchers to survive and prosper—as I soon learned when I happened upon a book that revealed to me the history of my profession.

In 2012, Mike's uncle passed away, and in his wake he left a treasure trove of old books. I adore books, especially old ones. Hunting through his bookcases, for me, was like spending a day at Tiffany's. One day, I came across a book that caught my eye. The book was called *The Crisis of the Old Order: 1919–1933 (The Age of Roosevelt, Volume I)*, by Arthur Schlesinger, Jr. I immediately snatched it up for my evening read. As I began to flip through the pages, I found that the most arresting part, in my opinion, was how reminiscent it was of America's current circumstances. The book could easily have been written today, rather than in 1925.

In 1925, Calvin Coolidge was President of the United States, and his belief was that America was now a business country that wanted a business government. This was a seismic shift from our former economy, which was more agriculturally based. During this same time, Andrew Mellon was the new Secretary of the Treasury. In 1925, Mellon was seventy years old, and in his lifetime, he had witnessed America's transformation into the greatest industrial nation in the world. This transformation dramatically affected the lives of all Americans, especially the farmer and the rancher.

It was a painful transition. The American farmer was suffering untold miseries. Their lives were filled with constant turmoil. Collectively, they were ready to revolt, having found no sympathy in Washington. The fact that farmers had recently lost their foreign markets due to the advent of World War I only exacerbated the rural depression unfolding across the nation. As a result, the average farmer lost a huge amount of income. Worse still, there was no place to go to replace that lost revenue.

Big business, on the other hand, was creating a contagious amount of excitement. The great industrial boom was applauded and supported in Washington. It was the political hot button. In this atmosphere, Coolidge remarked philosophically to the then chairman of the Farm Loan Board: "Farmers have never made money"—a simple statement, yet so bold in its connotations. Essentially, Coolidge's comments suggest that because the American farmer has always struggled, government doesn't need to worry about them. Farmers and ranchers will keep on doing what they are doing regardless of profit or loss.

In many ways, this sentiment has never really left us. Popular thinking is still that *this group of people has always suffered—so let's move on to something more exciting to talk about.*

Why should we care today about what was going on in 1925? The ideology prevalent during that time in history teaches us lessons about how the agricultural and ranching industries were perceived, and warns us how public opinion of ranching today has been negatively shaped by the past. Ranchers will face many difficult challenges in the years ahead, and we need positive public opinion to provide the support system necessary to face those challenges. Without it, the consumer will suffer too.

That suffering is looking more and more likely thanks to yet another problem facing small food producers: the disinterest of young people. In March of 2014, the *Rural Virginian* published an article about what they called "the last of the farmers." The article examined the biggest worry for most aging farmers: the fact that they are land-rich and cash-poor. Their farms are their 401K, and the only way to continue in the business if their children are not interested in carrying on their work is to bring in someone from outside the family to run their operations, or close down and sell their assets. Meanwhile, the younger generation doesn't want to live outside the city. They want convenience, and they aren't interested in the hard work and hardship involved in farming. Because farming/ranching is not very appealing to the younger generation, we are seeing more and more liquidation as young people refuse to go into their families' businesses.

But who could blame them? As the *Rural Virginian* article goes on to state, a new tractor can cost anywhere from $70,000 to $100,000. A new combine costs upward of $250,000. To engage in this type of business, you would typically have to invest $100,000 to make $10,000 a year in income. It doesn't take a Harvard graduate to see that the numbers are not very compelling.

Steve Forbes, the chairman and editor-in-chief of Forbes Media, recently wrote an article for the *Houston Chronicle* that was excerpted in *American Cattlemen* magazine, talking about today's agricultural challenges. In his article, Forbes contends that American agriculture and agribusiness as it operates today is "under assault in a way that will harm the ability of the world to feed itself." Today's ranchers and farmers are facing exploding business costs, sky-high land prices, crazy weather patterns, and the lowest cattle numbers this country has seen in fifty years.

Forbes goes on to say that science-driven agriculture needs more applause, rather than criticism. Today, American society is part agricultural, part industrial, and part technological. When integrated, these different sub-groups can form a strong whole. We can use industrialization and technology to help us address some of our biggest challenges—if we allow it.

If the farmers and the ranchers continue to bail, our food choices will become drastically limited. Yet one thing will never change: we all have to eat to live. It's a serious problem without an easy answer, but it is clear that, to build a profitable future, American farmers and ranchers have to learn from the past. We must assuage the fears of the younger generations and consumers alike. As an industry, we have to develop a voice stronger than our detractors'. Often, a consumer's only source of information comes from an Internet search, and that information is not always accurate. It is up to us to promote our own industry and share the facts behind it. Taking an active stance can help cattlemen like us improve our image and gain the broad public support we need in order to survive and prosper—a dream that is becoming more and more difficult for the average farmer.

Fortunately, community support of local farmers and ranchers can be a start in the right direction—and that is happening all over the country. For my husband and I, the fresh movement is comforting, even though the challenges we face going forward are numerous. But we are tough, and committed to the long-standing traditions of the American rancher—and I am confident our drive and perseverance will triumph in the end.

CHAPTER 11

Reinvention

Everyone has a story to tell. Mine is about reinvention. I kept the good parts of the old me and rolled them into a new me in order to survive under a different set of circumstances. When it comes to adding new elements to our lives, we are only limited by the scope of our dreams.

I am now a rancher, and I wrote this book to give my perspective from a woman's point of view, because when we think of ranching, we typically think of men. But women have always played a very important role in any ranching operation—though maybe not as vocal of a role. I decided to give women ranchers a voice—my own, unique to my particular circumstances, but a voice nonetheless.

In my journey, I've been most fascinated to see just how many people my age are starting new businesses that are totally unrelated to their past careers. We learn skills throughout our lives, in all kinds of ways, creating a knowledge base that we can then transfer into a myriad of avenues.

As we age, the Baby Boomer generation is staying more focused and more active than past generations. We are also realizing that we have to keep on working longer, and that we *want* to—only now, our new careers must be ones that we choose, ones that involve subjects about which we are passionate. If they are, then our jobs won't seem like work.

Proceeding in a new direction in mid-life creates an incredible energy that flows through our veins. This exhilaration keeps us young and vibrant.

We are a formidable generation, and we are not done yet. I feel I am just hitting my stride. Life is a beautiful dance with our own unique spirit, and age doesn't have to limit what we can accomplish in one lifetime.

I spent the first half of my life being taken care of by others. I didn't have to work very hard. I enjoyed a lot of perks and I was sheltered from the fears and difficulties of taking on risk and working to make a living. These worries were carried on the backs of others. I am now being taken care of in a different way: emotionally and lovingly—but now, I also have to stand on my own two feet. I have learned how to take risks, move out of my comfort zone, and live less large.

I work hard, my husband works hard, and at the end of the day, sometimes we are too tired even to eat. We just fall into bed and forget about it, but it is an honest day, an honest living, and I have grown exponentially from it. I have learned that nature can be your best companion or your most challenging opponent. When I cook a great steak or throw a hamburger on the grill, I beam with pride knowing that I have played a small role in this great industry we call ranching.

During one of his recent visits, my son-in-law Alexis accompanied me on a hike of our property to check fences. I thought he would find it interesting to see our operation. As we crossed a beautiful meadow, the cows and calves contentedly grazing all over the hillside and the sun shining down upon the glistening stream that bubbled through the pasture, Alexis looked at me and said, "I love your office." Amen!

In the end, I will always be a princess in spirit—but now, I am a princess rancher.

Mountain shot with the cows on our new ranch in Lexington

About the Author

Kelly G. Williams has had the unique opportunity to experience firsthand the journey that transformed a small Virginia company into one of the largest food distributors in America. Food has been at the core of her life for over thirty years. Her travels have taken her all over the world and created a rich tapestry of food experiences and cultures that have shaped her philosophies on cooking and food production. She currently raises Angus beef cattle with her husband in Lexington, Virginia. Visit her at www.princessrancher.com

www.ingramcontent.com/pod-product-compliance
Lightning Source LLC
Chambersburg PA
CBHW031212090426
42736CB00009B/880